China Forest Certification Development Report 2021

Science and Technology Development Center of National Forestry and Grassland Administration
Research Center for Forest Certification of National Forestry and Grassland Administration

中国林業出版社
CFPH China Forestry Publishing House

图书在版编目(CIP)数据

中国森林认证发展报告. 2021 = China Forest Certification Development Report 2021：英文／国家林业和草原局科技发展中心，国家林业和草原局森林认证研究中心主编. —北京：中国林业出版社，2022. 3

ISBN 978-7-5219-1567-9

Ⅰ. ①中… Ⅱ. ①国… ②国… Ⅲ. ①森林经营-认证-研究报告-中国-2021-英文 Ⅳ. ①S75

中国版本图书馆 CIP 数据核字(2022)第 021686 号

出版 中国林业出版社(100009 北京西城区刘海胡同 7 号)
电话 010－83143564
发行 中国林业出版社
印刷 北京中科印刷有限公司
版次 2022 年 3 月第 1 版
印次 2022 年 3 月第 1 次
开本 787mm×1092mm，1/16
印张 4. 75
字数 150 千字
定价 68. 00 元

PREFACE

Forest certification, as an effective mechanism in promoting sustainable forest management and market accessibility of forest products, is widely recognized by the international community. Chinese government attaches great importance to sustainable forest management and forest certification. Since 2001, by combining the international practices with Chinese situations, China Forest Certification Council (CFCC) has been established, and a unified national forest certification scheme has been built, which was mutually recognized with PEFC in 2014. CFCC covers forest management, chain of custody, non-timber forest products, bamboo forest, forest ecosystem services, carbon neutrality, etc. By years of implementation, forest certification has been proven to become an effective means in enhancing quality development of forest in the new era.

As the reform and opening-up and ecological civilization construction advance continuously, China enters into a brand new development phase, in which it implements the new development concept featuring innovation, coordination, green, opening-up and sharing in an overall manner, and strives to build the new development pattern with the domestic cycle as the main body and international cycle as the propeller. The new context created new opportunity for forest certification. The amendment of the *Forest Law of the People's Republic of China* has come into effect on July 1, 2020, with which clearly stipulating "Forest managers can apply for forest certification on a voluntary basis, to improve their forest management practice and to promote sustainable management" , provides the legal basis for comprehensive promotion of forest certification in China.

To fully demonstrate the progresses of China's forest certification to the international community, and contribute the Chinese experiences to the global forest certification, the National Forestry and Grassland Administration organized an expert team to write the *China Forest Certification Development Report 2021*. The report introduces the overall progresses and achievements in different phases of China forest certification, which include the perspectives of development, system construction, technical standards and practical experiences. It will provide valuable references and information for foreign and domestic organizations, personnel, professionals and managers in relevant fields who wish to dedicate to forest certification research, management and practice.

CONTENTS

CHAPTER I
Overview of China Forest Certification

1.1 Overview of China forestry

Comparing to the backdrop of the overall downward trend of global forest resources, based on its own national forest conditions, China has actively pursued the basic policy of ecological priority and green development. It took full advantage of its social system and mobilized all level social sectors to join the forestry development. During the past 30 years, it has continuously realized the double growths in forest area and stock, and made remarkable achievements in improving ecological environment and forestry development.

1.1.1 Status of forest resources

At present, the forest area nationwide has reached 220 million ha, of which the plantation forest area has reached 80.0 million ha. The forest coverage increases to 23.04%, and forest stock exceeds 17.5 billion m^3. The average forest area per capita is 0.16 ha, and the average forest stock is 12.35 m^3. The total biomass of forests across the country is 18.36 billion tons, and total carbon stock reaches up to 9.19 billion tons. Annually, the forest conserves 628.95 billion m^3 of water, fixes 8.75 billion tons of soils, saves 462 million tons of fertilizers, absorbs 400 million tons of pollutants, and retains 6.16 billion tons of dusts.

1.1.2 Dynamic changes of forest resources

According to China's National Forest Resources Inventory, the forest coverage in China increased from 18.21% (1999) to 22.96% (2018); Forest area steadily increased, with the net forest area increased by 455.54 million ha and net forest stock increased by 5.10 billion m^3. From 2004 to 2018, the net plantation forest area increased by 18.34 million ha, and net plantation forest stock increased by 1.43 billion m^3, mean while, net natural forest area increased by 20.72 million ha, and net natural forest stock increased by 2.27 billion m^3.

The changes of forest resources in China mainly showed the following characteristics: ① The forest area increased steadily, while forest stock increased rapidly; ② The forest structure has improved, mid-aged, young-aged and low-density forests proportionally reduced, and forest quality continuously increased; ③ The forest logging has decreased, with average forest stock annually increasing, and forest resource surplus continuously widening; ④ The supply capacity of commercial forest has improved, and the ecological function of public welfare forest has enhanced; ⑤ The steady increase of plantation forest has become an important guarantee to maintain the stability and sustainable growth of forest area, while natural forest still plays an important role in forest restoration; ⑥ The proportion of forestland area managed by individuals has increased significantly, and the reform of collective forest tenure has made remarkable achievements.

1.1.3 Production and trade of forest products

Up to 2018, forestry industry had been keeping growth in a largescale and medium-high speed. Nationwide, the production value of the primary industry of forestry was RMB 250 million yuan, the secondary industry RMB 350 million yuan, and the tertiary industry RMB 170 million yuan. With the production value ratio of the primary, secondary and tertiary industry being 33:45:22, the forestry industrial structure gradually improved. Since the reform and opening-up, China's international cooperation in forest products has furtherly developed, with rapid increase in export and import. In 1978, the total import and export of forest products were RMB 35.5 billion yuan, and the forestry GDP was RMB 367.81 billion yuan; By 2018, the total import and export of forest products, the forestry GDP reached up to RMB 1.04 trillion yuan and RMB 7.63 trillion yuan respectively(On the basis of current price).

1.2 Law and policy environment of forest certification

In recent years, China has successively issued a number of laws and policies related to forest certification and forestry industry, which forcefully pushed China forest certification forward. Currently, a basic law and policy environment supporting the healthy development of forest certification has been created.

1.2.1 Laws

1.2.1.1 Forest Law of the People's Republic of China

In 2019, President Xi Jinping signed No.39 the Order of the President of the People's Republic of China , to publish the amended *Forest Law of the P.R.C.*, which was approved by the 15th Session of the Standing Committee of the 13th National People's Congress. According to the Article 64 of this law, "Forest managers can voluntarily apply for forest certification, in order to improve their forest management practice and to promote sustainable management." The provision mentioned above demonstrates that forest certification has been officially incorporated into the national legislation in China.

1.2.1.2 Government Procurement Law of the People's Republic of China

In 2002, Chinese government officially published the *Government Procurement Law of the P.R.C.*, to introduce the policies on public bidding and public procurement of some public products. Because the types and quantity of certified products are not enough to provide sufficient products supply. The current government procurement policy has not yet covered forest certified products. However,if the conditions are right in the future and once the certified forest products would be listed as the priority in public procurement, it will greatly promote the domestic market of forest products certification.

1.2.2 Policies

1.2.2.1 Decision of the State Council in Accelerating the Forestry Development

In 2003, the State Council and Central Committee of the Communist Party of China jointly published the *Decision on Accelerating Forestry Development*, which specifies to deepen the forestry system reform, accelerate the establishment of the certification system for public–welfare forestry, boost the momentum for the development of forestry, actively promote the forest certification to align with the international practice, constantly optimize the forestry structure, and promote the forestry industry's development.

1.2.2.2 Rules for Forest Certification

In 2015, the Certification and Accreditation Administration of the P.R.C. and the State Forestry Administration jointly published the *Rules for Forest Certification*, which specifies to centralize management of forest certification, defines 7 certification scopes and corresponding certification bases for forest management, chain of custody, non–timber forest product management, bamboo forest management, eco–services of forest in nature reserve, eco–services of forest in forest park, wildlife husbandry and management, etc., and provides requirements for certification body, qualification and continued education of certification auditors, certification procedure, certificate, certification information report, certification body accreditation and registration of certification personnel. Through the implementation of a series of technical specification, the forest certification bodies and certification auditors were comprehensively and effectively managed and regulated.

1.2.2.3 Action Plan on Forestry Brand Building and Protection 2017-2020

In 2017, the State Forestry Administration published the *Action Plan on Forestry Brand Building and Protection 2017–2020*, which specifies to strengthen forest certification, improve the forest certification system, expand the certification scope, facilitate the certified forest products to be included into the list of government procurement, and promote the forest certification in order to improve the social recognition for certified products.

1.2.2.4 Notice on Regulating the Operating of Forest Certification Practice

In 2018, the State Forestry Administration published the *Notice on Regulating the Operating of Forest Certification Practice*, which further emphasizes to thoroughly implement the forest certification scheme, clarifies that forest certification is centrally managed by the Ministerial Joint Conference for Certification and Accreditation, encourages the participation and implementation of forest certification by law, concrete implementation of forest certification practice, and proactively accelerates the internationalization of forest certification. Meanwhile, it clearly requires the provincial-level forestry authorities to raise awareness, incorporate forest certification into their important agenda, develop local policies, promote the trust to forest certification result in an orderly manner, so as to create a favorable environment for forest certification development.

1.2.2.5 Strategy on Natural Forest Conservation and Restoration

In 2019, the State Council published the *Strategy on Natural Forest Conservation and Restoration*, which provides to establish a natural forest conservation and restoration system featuring comprehensive conservation, systematic restoration, use management and control as well as clearly-defined responsibilities, to ensure gradual increase of natural forest area, continuous quality improvement, and steady function progress. It helps to give full play to the pilot forest certification of *Natural Forest Conservation Project* in playing a leading and demonstrating role.

1.2.3 Guidelines

1.2.3.1 Opinions on Conducting Forest Certification

In 2008, the Certification and Accreditation Administration of the P.R.C. and the State Forestry Administration jointly published the *Opinions on Conducting Forest Certification*, which clearly states that China's forest certification scheme belongs to the national uniform voluntary certification system, encourages the trust in China forest certification results, and strengthens the guidance and monitoring over forest certification.

1.2.3.2 Guiding Opinions on Promoting Forest Certification

In 2010, the State Forestry Administration published the *Guiding Opinions on Promoting Forest Certification*, which points out that forest certification is required by modern forestry, and it will help the forest resources' transformation from direct government management to joint supervising and management by government and society, forest management's transformation from economic benefit-oriented to balanced and comprehensive ecological, social and economic benefits, forest logging's transformation from quota-based management to forest management-based, timber transportation's transformation from license-based to certification label-based, and forest products utilization's transformation from excessive consumption to green consumption.

1.2.3.3 Guiding Opinions of the State Council on Accelerating the Establishment and Improvement of Green, Low-carbon and Circular Economic System

In 2021, the State Council published the *Guiding Opinions of the State Council on Accelerating the Establishment and Improvement of Green, Low-carbon and Circular Economic System*, which specifics to complete and improve the production system for green, low-carbon and circular development, develop the forestry's circular economy, push products' green design, build green manufacturing system, develop remanufacturing industry, and strengthen certification, promotion and application of remanufactured products. Meanwhile, it clearly provides to improve a distribution system for green and low-carbon circular development, develop high-value-added quality green products trade, reinforce international cooperation in green standard, actively lead and join the development of relevant international standard, boost the cooperation in conformity assessment and mutual recognition, and align the green trade rules with the import and export policies to facilitate the cooperation under the green "Belt and Road Initiative."

1.2.3.4 Guiding Opinions of the National Forestry and Grassland Administration on Promoting Quality Development of Forestry and Grassland Industries

In 2019, the National Forestry and Grassland Administration (NFGA) published the *Guiding Opinions of the National Forestry and Grassland Administration on Promoting Quality Development of Forestry and Grassland Industries*, which requires to properly handle the relationship of forest and grassland resources conservation, cultivation and utilization, build an ecological industry system of forestry and grassland featuring with ecological construction in the way of industrialization and industrial development by ecological route, so as to effectively enhance national ecological security and timber security, and boost rural vitalization, poverty alleviation as well as social and economic development.

1.2.3.5 Opinions on Establishing and Improving Value Realization Mechanism of Ecological Products

In 2021, the State Council and Central Committee of the Communist Party of China jointly published the *Opinions on Establishing and Improving Value Realization Mechanism of Ecological Products,* which clearly requires to build the ecological products' investigation and monitoring mechanism, value assessment mechanism, operation and development mechanism, protection and compensation mechanism, value realization supporting mechanism, and value realization promotion mechanism, in order to implement the idea "Lucid waters and lush mountains are invaluable assets" . It requires to actively providing more quality ecological products to meet people's increasing demand for beautiful eco-environment, cultivate the new business type and new business model for transformation to green development, and build a favorable eco-environment as a powerful support for continuous healthy economic and social development.

1.3 Development of forest certification

China's forest certification scheme from the introduction of the concept to the present comprehensive establishment can be divided into the initial stage, the construction stage and the development stage.

1.3.1 Initial stage (2001-2008)

In 2001, the establishment of China's Forest Certification Scheme was officially initiated. The China Forest Certification Working Group was established, and the Forest Certification Division was set up in the Scientific and Technological Development Center, State Forestry Administration, to take charge of building the China's Forest Certification Scheme as well as planning, implementing and managing forest certification activities. In 2002, the development of forest certification standards and capacity building were started. In 2003, the *Decision of the State Council on Accelerating the Forestry Development* provides to "actively carry out forest certification in order to gear to international practices." In 2005, the pilot projects of forest certification were initiated. In 2006, the pilot work was introduced into the first batch of 6 provinces, including Jilin, Heilongjiang, Zhejiang, Fujian, Guangdong and Sichuan. In 2007, the forestry standards for forest management certification and chain of custody certification were published, marking the first forest certification standards developed and published in China. In 2008, the Certification and Accreditation Administration of the P.R.C. and the State Forestry Administration jointly published the *Opinions on Implementing Forest Certification*, and set up the National Technical Committee for Sustainable Forest Management and Forest Certification Standardization.

1.3.2 Building Stage (2009-2014)

In 2009, the *Implementation Rules of China Forest Certification* was published, and Zhonglintianhe Certification Center, the first certification body endowed with the forest certification qualification in China, was formally established. In 2010, the Forest Certification Leading Group was set up in the State Forestry Administration, China Forest Certification Committee was established, and the first training for forest certification auditors was provided. In 2011, China Forest Certification Committee officially became a national member of the Program for the Endorsement of Forest Certification Scheme (PEFC), the pilot projects of collective forest certification were implemented, and the website of China Forest Certification Council was launched. In 2012, the national standards for forest management certification and chain of custody certification were published, and the application and documents for mutual recognition were officially submitted to PEFC Secretariat. In 2013, the First Meeting of Forest Certification Stakeholders Forum was held. In 2014, China Forest Certification Council (CFCC) and PEFC realized mutual recognition, and the pilot of forest certification system was officially kicked off.

1.3.3 Development Stage (2015 up to present)

In 2015, The *Rules for Forest Certification, the Rules for Accreditation of Forest Certification Body and the Rules for Registration of Auditors* were published, forming the framework of forest certification scheme. In 2016, the first workshop for forest certification bodies was organized since the uniform forest certification scheme had been established, and CFCC attended the international forest certification workshop and the exhibition. In 2017, the Certification and Accreditation Administration of the P.R.C and the State Forestry Administration jointly published a notice to include the forest certification in "planted endangered plants" into the *Rules for Forest Certification*, improved the examination outline for forest certification auditors, established the China Forest Certification Promotion Department in East China and the Promotion Center in South China, officially initiated the first "China Forest Certification—Chain of Custody Demonstration (Pilot) Projects" in the timber and bamboo industries, published the serial standards for green products assessment, and incorporated the forest certification requirements into the green products assessment indicators for furniture, wood-based panel and wood flooring. In 2018, the forest certification workshop was organized, and 4 forestry standards for forest certifications in carbon-neutral products, forest fire brigade construction, wildlife feeding management and label use requirements were published. In 2020, the second mutual recognition with PEFC was initiated, the standard of China Forest Certification for Non-timber Forest Products was published, and the 4 forestry standards concerning natural reserve, forest wellness, etc. were issued.

After nearly 20-year continuous efforts, China forest certification has made great achievements. The CFCC which is accordance with the international principles and Chinese national conditions and forest situation was well established, the certified forest area and certified products were continuously increased, the standards system was increasingly improved, capacity building was strengthened, significant ecological, economic and social benefits were realized, and the international impacts of certification scheme were increasingly intensified. Up to May 2021, China has published 32 forest certification standards, and issued forest management (FM) certificates for 5.8 million ha of forests and chain of custody (CoC) certificates for 361 timber processing, manufacturing and trading companies. Besides, China certified 15 wildlife husbandry and management companies, 62 non–timber forest products enterprises, 6 rare endangered plants managers, and 2 forest eco–environment service providers.

CHAPTER Ⅱ
China's Forest Certification Scheme

2.1 Organization structure

China's forest certification scheme consists of 4 parts, China Forest Certification Council, National Technical Committee of Sustainable Forest Management and Forest Certification Standardization, Forest Certification Research Center and Forest Certification Promotion Agency. China Forest Certification Council is the decision-making organization in the forest certification scheme; National Technical Committee of Sustainable Forest Management and Forest Certification Standardization is responsible for the standard development and revision; Forest Certification Research Center is responsible for all forest certification related science researches; and Forest Certification Promotion Agency is responsible for the publicity and marketing of China's forest certification scheme and certified forest products. See the organizational structure of China's Forest Certification Scheme in Fig. 1.

2.1.1 China Forest Certification Council (CFCC)

In November 2011, China Forest Certification Council (CFCC) was officially established. CFCC consists of members from government departments, research institutions, universities, wood-based products producers and manufacturers as well as social societies. Mainly, it is responsible for the drafting, final review and publication of technical documents concerning China's forest certification scheme, operation and technical support for the system, resolution of disputes, complaints and grievances concerning the system, its publicity and promotion, and international communication and cooperation on behalf of the scheme. CFCC has a Secretariat, a Stakeholders Forum and a Dispute Resolution Committee. The Stakeholders Forum provides a communication platform for all stakeholders who are interested in the development of China forest certification. The Dispute Resolution Committee takes charge of disputes resolution in operating China's forest certification scheme, to ensure the objectivity, impartiality and effectiveness of forest certification, and to secure and reflect the rights and reasonable demands from the forest certification applicants, certification bodies, accreditation bodies and stakeholders.

2.1.2 National Technical Committee of Sustainable Forest Management and Forest Certification Standardization (TC360)

In April 2008, the Standardization Administration of China approved to build the National Technical Committee of Sustainable Forest Management and Forest Certification Standardization (TC 360, hereinafter referred to as the Technical Committee), which takes charge of organizing the development and management of forest certification standards and technical specifications, and providing technical support for forest certification activities. The Technical Committee consists of representatives from government departments, research institutions, industrial associations, education institutions and enterprises.

2.1.3 Forest Certification Research Center

To boost the construction of China's forest certification scheme, the "Forest Certification Research Center of State Forestry of Administration" was set up in the Chinese Academy of Forestry in 2013, to be responsible for theoretic and policy research of forest certification, development of relevant standards and technical specifications, forest certification training, consultation, demonstration and promotion, international communication, promotion of mutual recognition with international certification, establishment of forest certification information platform, and management of China forest certification trademark. Its foundation was of great importance to the integrated development of forest certification research, demonstration and promotion, and improvement of Chinese scheme's competitiveness on international market.

2.1.4 Forest Certification Promotion Agency

In 2017, the CFCC Promotion Department in East China and the CFCC Promotion Center in South China were established in Jiangsu and Guangdong respectively, to meet the actual needs of promotion and application of forest certification. The promotion agencies help to increase the public awareness to forest certification, to facilitate the production and management companies, marketing companies and consumers to fully accept the concept of sustainable management, and further boost the development of certification market.

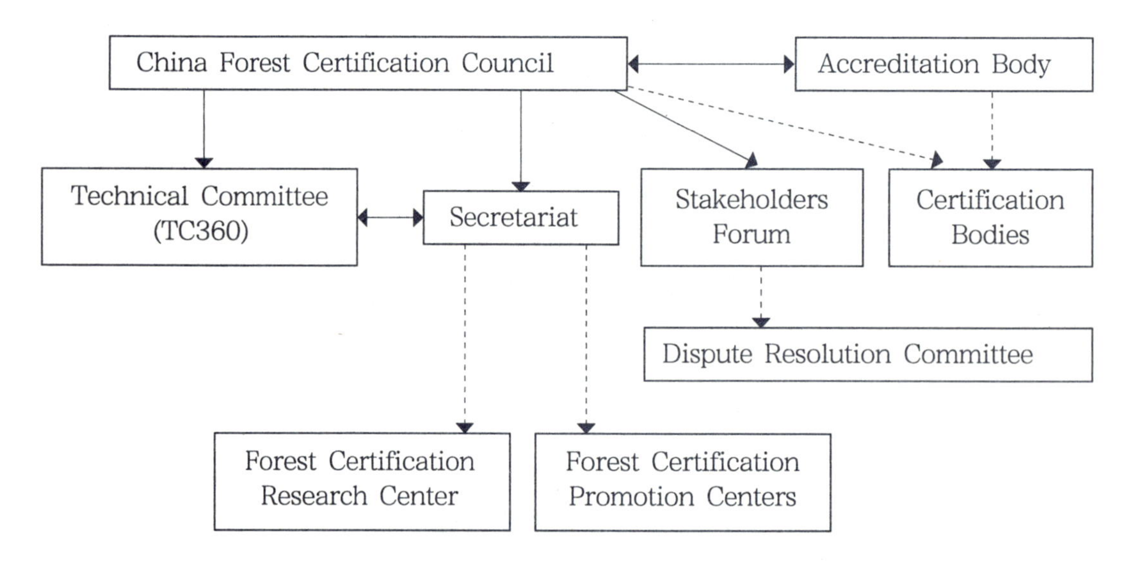

Fig. 1 Organization Structure of China's Forest Certification Scheme

2.2 Standards system

In the CFCC standards system, there are 32 standards, including 3 national standards and 29 forestry standards. The 3 national standards include the *Forest Management* (GB/T 28951–2021), *Chain of Custody* (GB/T 28952–2018), and *Non–timber Forest Products Management* (GB/T 39358–2020).

The CFCC standards system covers 7 categories, namely basic and general requirements, forest management, ecosystem service, wildlife management, chain of custody and carbon neutrality (See Table1).

Table 1 CFCC Standards System

Category	Standard No.	Standard Name
Basic and General Requirements		Terminology
		Standard Development Rule
		Procedures for Use of Certification Trademark
		Group Forest Certification Requirements
		Capacity Requirements for Certification Bodies and Personnel
Forest Management	GB/T 28951–2021	Forest Management Requirements
		Application Guidance for Forest Management Standard
	GB/T 39358–2020	Non–timber Forest Products Management Requirement
		Application Guidance for Non–timber Forest Products Management Standard
		Bamboo Forest Management Requirement
		Application Guidance for Bamboo Forest Management Standard

Cont.

Category	Standard No.	Standard Name
Ecosystem Service	LY/T 2239-2013	Eco-environment Services Requirements in Natural Reserve
	LY/T 2277-2014	Eco-environment Services Requirements in Forest Park
		Resource Management Requirements in Natural Protected Area
	LY/T 3246-2020	Eco-tourism Requirements in Natural Protected Area
	LY/T 3245-2020	Forest Wellness Requirements in Natural Protected Area
Wildlife Management	LY/T 2279-2019	Wildlife Husbandry and Management Requirements
		Application Guidance for Wildlife Husbandry and Management Standard
	LY/T 2602-2016	Rare and Endangered Plants Management Requirements
		Application Guidance for Rare and Endangered Plants Management Standard
Chain of Custody	GB/T 28952-2018	Chain of Custody Requirements
		Application Guidance for Chain of Custody Standard
Carbon Neutrality		Forest Carbon Sink Requirements
	LY/T 3116-2019	Carbon Neutral Products Requirements
Others	LY/T 3117-2019	Forest Fire Brigade Construction Requirements
		Flower Management Requirements
		Biohazard Prevention and Control Management Requirements
		Grassland Management Requirements
		Wetland Management Requirements

2.3 Certification type

In China's forest certification scheme, there are forest management certification, chain of custody certification, non-timber forest product certification, bamboo forest management certification, forest eco-environment service certification, rare and endangered plants management certification, etc. In the future, the certification will cover natural protected area, grassland and wetland according to development needs.

2.3.1 Forest Management Certification

Forest management (FM) certification is issued for forest management units. It demonstrates that the forest management unit has been managing the forest in a socially beneficial, environment–friendly and economically viable manner.

2.3.2 Chain of Custody Certification

Chain of custody (CoC) certification is to allow organizations to provide accurate and verifiable information that forest and tree based products are sourced from CFCC certified sustainably managed forests, recycled material and controlled sources. Through tracking products from forest and throughout the whole supply chain, the CoC certification ensures the traceability of product material origin.

2.3.3 Non–timber Forest Products Certification

Non–timber forest products (NTFP) refer to products other than timbers from the forest, including plant fruits, resin, latex, essential oil, fiber, feed, fungi, etc. NTFP certification is to assess the NTFP's production and management activities, including cultivation, harvest, storage, transportation, sales, etc., and to testify whether it has practiced sustainably.

2.3.4 Bamboo Forest Management Certification

Bamboo forest management certification is to audit and assess the operation and management of the forest management units with bamboo forest as the main species. Bamboo forest management certification can effectively reduce the negative environmental, economic and social impacts on bamboo forest management.

2.3.5 Forest Eco-environment Service Certification

Forest eco-environment service certification includes the forest eco-environment service certifications for forest park and natural reserve.

Forest park eco-environment service certification is to audit and assess the eco-environment service function of a forest park. It can promote the sustainable forest park management, and fully develop the forest parks' multiple values in resource conservation, biodiversity, ecological culture dissemination and eco-tourism.

Natural reserve forest eco-environment service certification is to certify the eco-environment service function of a natural reserve. It can improve the natural reserve management units' practice, and natural reserves' eco-environment service function, so as to realize its sustainable development.

2.3.6 Rare and Endangered Plants Management Certification

Rare and endangered plants management certification includes wildlife husbandry and management certification, and rare and endangered plants management certification.

Wildlife husbandry and management certification is to certify the human resource, infrastructure, feeding input, breeding, animal health, animal slaughter and product acquisition in the wildlife husbandry and management organization.

Rare and endangered plants management certification is to certify the artificial rare and endangered plants management units for production and operation purposes.

2.4 Accreditation and mutual recognition with international system

2.4.1 Accreditation

2.4.1.1 Accreditation body

National accreditation body is responsible for accrediting the forest certification bodies. The forest certification bodies in China need to be accredited by the national accreditation body to gain the qualification to carry out forest certification.

The *Accreditation Procedures of Forest Certification Body* (CNAS–SC23) provides the specific requirements and guidance for the accreditation of forest certification body, and serves as the accreditation basis together with the other accreditation rules of CNAS.

2.4.1.2 Accreditation requirements

Forest certification bodies can only be established upon the approval of the certification and accreditation monitoring and management department in the State Council, and can only start the certification practice within the approved scope after registered as a legal entity. A certification body can be established, when it meets the following requirements: First, it has a fixed office and necessary facilities; Second, it has adopted the management system which meets the certification and accreditation requirements; Third, it has the registered capital no less than RMB 3 million yuan; And fourth, it has over 10 full–time certification employees working in corresponding fields. Currently, there are 11 officially accredited forest certification bodies in China.

2.4.1.3 Auditor

At present, forest certification auditors need to be registered in China Certification and Accreditation Association (CCAA). The *Rules for Registration of Management System Auditors* stipulates the specific requirements for the registration of forest certification auditor, e.g. auditors are required to have the professional higher education background, relevant work experiences as well as corresponding expertise and skills.

2.4.2 Mutual recognition with international system

In August 2011, CFCC officially became a member of PEFC. In 2012, CFCC formally submitted the application for mutual recognition to PEFC. In February 2014, CFCC finally realized mutual recognition with PEFC.

CHAPTER Ⅲ
Development of China Forest Certification

3.1 Forest Management Certification

Since 2007, China's forest management certification has experienced the periods of slow growth, rapid increase and dynamic development. Currently, it is still in the continuously improving stage. With its significant ecological, social and economic benefits, China's forest management certification greatly boosted the sustainable forest management.

3.1.1 Certified area

In 2007, the forestry standard *Forest Management* (LY/T 1714–2007) was published and implemented. In 2009, the first forest management certificate was issued for the 30,000 ha forest in the state-owned Fushun County Forest Farm in Liaoning Province. From 2009 to 2020, the CFCC certified forest area reached 8.10 million ha, accounting for 3.7% of the total 220 million ha forests across China.

According to the incomplete statistics from CFCC website, 5.42 million ha forests have obtained CFCC forest management certificates by the end of 2020. See the dynamic changes of certified areas by year in Fig. 2.

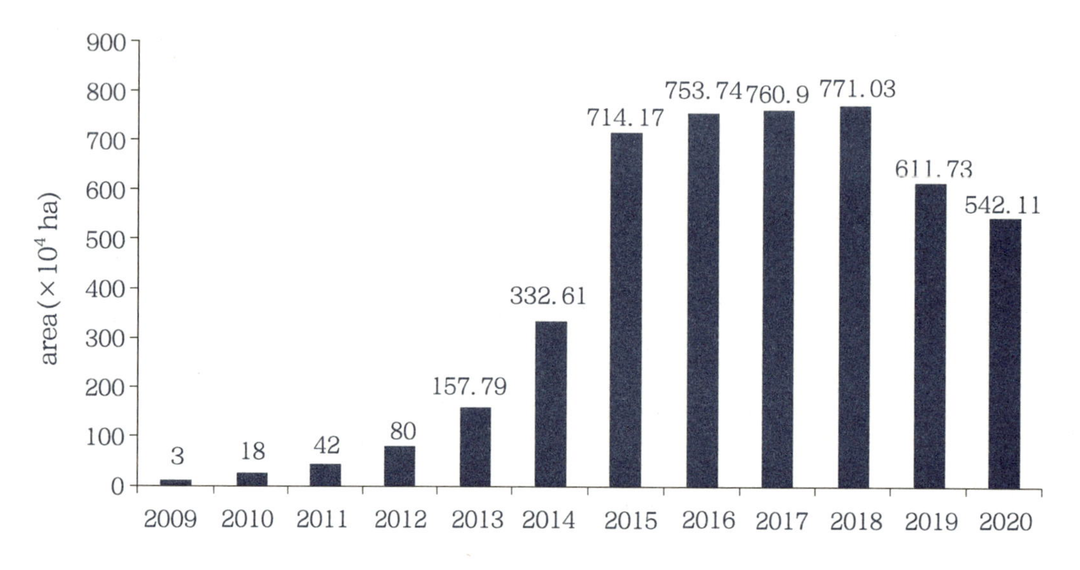

Fig. 2 Changes of certified forest areas in China (2009–2020)

The certified forest area can be as small as 7 ha, and can also be as large as 920,000 ha. FMUs under 1,000 ha accounted for 1/3, and those under 10,000 ha took up 60%.

Geographically, the certified forests located in 17 provinces, of which Heilongjiang, Guangxi, Yunnan and Qinghai have over 100,000 ha forests been certified, and Heilongjiang has the largest certified area, accounting for 87% of the total certified forests in China.

3.1.2 Certification benefits

Follow–up studies have shown that forest certification could generate multi–dimensional positive impacts on forest management.

3.1.2.1 Ecological benefits

Having been certified, the FMUs could improve their awareness and practices of eco–environment protection, optimize operational techniques, reduce forest management's negative impact, enhance forest's eco–functions, and realize the harmony and successes in production and ecological improvement.

By optimizing forest management activities, e.g. by using the cable skidding, setting up frames to attract eagles to staying, biodiversity conservation was greatly improved. In the forest area managed by Linjiang Forestry Bureau in Jilin Province, wildlife significantly increased in number, and species like wild boar, roe deer, pheasant and woodpecker became very commonly seen. In the state-owned Mulan Forest Farm in Hebei Province, the cable skidding was used to minimize damages to barks, shrubs and grass, and consequently the nature regeneration seedling quantity of precious tree species such as *Phellodenedron amurense, Fraxinus mandshurica, Tilia amurensis, Juglans mandshurica*, etc. increased obviously.

3.1.2.2 Social benefits

By carrying out forest certification, forest management units pay more attention to the interests of stakeholders, improve the awareness of responsibility to workers, surrounding communities residents, hired labor workers and other different stakeholders, and promote the interests of all parties were effectively secured. For example, through developing non-timber forest product management, the employees' and neighboring farmers' economic income was increased; by prioritizing the neighboring residents in employment, the villagers got more job opportunities and more sources of income; and through creating forest landscape and improving the quality of scenic spots, the consumers' demand for recreational ecological tourism was satisfied.

3.1.2.3 Economic benefits

With the forest certification, the FMUs got to improve their product quality, and expand their market access. The increased product price brought greater economic benefit to the FMUs. For example, the certified state-owned forest farms could sell their timbers at a price of RMB 10–20 yuan/m^3 higher than the previous one in average, and they could increase the production of non–timber forest products as pine nuts, wild vegetables and bee honey, generating more values from forest products.

Case 1

APP's Achievements with Forest Certification

In 2010, 5 subsidiaries of APP (China), invested by APP (China) Investment in mainland China obtained its forest management certificate for the 186,000 ha forests. With the certification practice, it significantly improved the forest management. Firstly, it manages the production and operations in nursery, afforestation and logging steps in a more normalized and standardized way. Secondly, it continuously improved its operation and management practices, increased the managers' and employees' awareness to occupational health and safety, and made great progresses in workers' rights and interests protection as well as compliance with forest laws and regulations. Thirdly, it got access to stable and more reliable sources of certified materials. And thus it reduced its reliance on imported certified materials, and improved the customers' satisfaction, fundamentally securing its access to international market. With the added value of RMB 100 yuan per ton, it could sell the certified products at higher prices. Fourthly, the APP's corporate image and product awareness were improved all–roundly, which not only generated a favorable demonstration effect for the China's paper–making industry, but also boosted the international trade organization's confidence in the transparency and integrity of China's forestry system.

Case 2

Daxing'anling Forestry Group's Achievements with Forest Certification

In 2014, 7 forest management units under the Daxing'anling Forestry Group obtained the forest management certificates for their 5.3 million ha forests. Through certification, it significantly improved its forest management and non-timber forest products management, increased the public awareness of forest resource protection, and made the concept of sustainable forest management widely accepted. Meanwhile, the certified products generated great economic benefits. For example, the Daxing'anling Songyi Foodstuffs Co., Ltd. exported its certified products to Singapore, Malaysia, the U.S., Korea and Japan; And Daxing'anling Luyuan Bee Products sold its certified forest products to international markets with a 15% premium.

3.2 Non-timber Forest Product Certification

China is rich in non-timber forest product resources. Chinese government attaches great importance to the sustainable management of NTFPs, and adopted the certification as one of the important measures to realize the sustainable management of NTFPs.

3.2.1 Overview of certification

For a long time, the forestry authorities mainly focused on timber and timber products in China. But non-timber forest products now attracted wide attentions with their unique characteristics, such as their short harvest time, high relevance to the local people's livelihood, higher unit price in contrast with timber, and their increasingly important role in improving the forest communities' income, providing job opportunities and eliminating poverty. In the context, how to ensure the sustainable management and production of NTFPs became an issue to be addressed. By introducing the forest certification mechanism into the NTFPs management, China provided an innovative solution.

NTFP certification is an integral part of China's forest certification scheme. It is to assess the FMUs' production and management process of non–timber forest products by the third–party certification body according to the NTFP certification standard as well as the certification procedures and requirements, including NTFP cultivation, harvest, storage, processing, packaging, transportation and sales, so as to demonstrate that the products are originated from sustainably managed forest or forest environment. NTFP certification covers all plants and fungi obtained from forest environment other than timbers and bamboo timber.

3.2.2 Certification standard

CFCC NTFP certification adopts the *Non–timber Forest Product Management* (GB/T 39358–2020), which was officially implemented on June 1,2021. The standard provides a series of requirements for carrying out NTFPs management activities in forest environment under the premise of meeting requirements of certification standard on forest management. The standard mainly includes 5 parts.

3.2.2.1 Basic requirements

Basic requirements cover 4 parts, lawful obligation, operation management, rights and interest protection as well as production activities. It requires that the NTFP management must be conducted in compliance with the existing national laws and regulations, and by following the right procedures. Meanwhile, it requires fulfill obligations and responsibilities in accordance with the international conventions.

3.2.2.2 NTFP management plan

NTFP management plan includes three aspects, i.e. planning, review and publication, implementation and revision of the NTFP management plan. It requires reasonable utilization of human, material and financial resources, and proper organization of NTFP management activities, on the basis of considering the changes in NTFP management goal, market demand, internal and external environment.

3.2.2.3 NTFP management activities

NTFP management covers 3 parts, development and implementation of cultivation technologies, technical requirements for harvest, and wild plant harvest. It stipulates to standardize the production and operation activities, including NTFP cultivation, harvest, storage, transportation and sales, by using the published technical specifications for management, or by developing corresponding technical documents for operation and management.

3.2.2.4 Forest environment protection

Forest environment protection covers 6 tasks, namely water and soil resource conservation, biodiversity conservation, use of chemicals, disposal of wastes and pollutants, forest protection and environmental impact assessment. It requires effective conservation of forest resources and forest eco-environment in the process of NTFP production and management.

3.2.2.5 Management monitoring and documentation management

This part covers 2 tasks, production and sales management, and documentation management. It requires regular monitoring of NTFP production and management, and systematic management of records.

3.2.3 Comparison between NTFP certification and other certifications

At present, there are multiple NTFP certifications in China. Besides the CFCC NTFP certification, there are organic products certification, geographical indication product certification and sustainable agriculture network standard certification. The 4 certifications are compared in Table 2 to differentiate them from CFCC NTFP certification.

Table 2 Comparison of 4 NTFP certifications

Item	NTFP	Organic Product	Geographical Indication Product	Sustainable Agriculture Network
Intention	To promote the sustainable management of non-timber forest product resources	To boost the development of organic agriculture	To protect products with special qualities in special environment, and to promote local development	To eliminate the environmental and social risks caused by production through incentivizing continuous improvement
Basis	Including 2 standards: Forest management certification standard and non-timber forest product management certification standard	Including 4 national standards, which cover 4 parts: Production, processing, logo, sales and management system	General standard requirements	Including standard and appendix
Scope	Covers the whole process management throughout the production and operation procedures, from cultivation to harvest, storage, processing, packaging, transportation and sales	Covers the whole production process of organic plants, animals and microbiological products	Covers products with special quality, reputation and other characters from special geographical environment	Covers the whole farm, infrastructure, processing and packing area, protection and living area as well as individuals influenced by production activities
Applicability	Applicable to plant and fungi products from forest environment, other than timber (or bamboo timber)	Applicable to the production, harvest, post-harvest processing, packaging, storage and transportation of organic plant, animal and microbiological products	Applicable to geographically indicated products accredited by national administrative authority	Applicable to all farm types
Product mix	Primary and processed products	Primary and preliminarily processed plant, animal and microbiological products	Primary food products	Primary products

Cont.

Item	NTFP	Organic Product	Geographical Indication Product	Sustainable Agriculture Network
Technical requirements	To pay attention to the performance requirements for operation, and meet the social, environmental and economic sustainability requirements	Not to use chemicals and GM technology	Special geographical environment and special quality	Production should meet the social, environmental and economic sustainability requirements
Internation-alization	Mutually recognized with PEFC	Corresponding international standard available	Corresponding international standard available	International standard

3.2.4 Status of certification

In 2014, the State Forestry Administration published the *Notice of the SFA on Pilot Certification for Non-timber Forest Product*, which boosted the rapid development of NTFP certification in China. The certified area gradually progressed from the state-owned forest area in North China to the forest area in North China Plain, and then to the collective forest area in the Southern part. The certified NTFP types also increased from the 8 major types, nuts, berries, fungi, wild vegetables, bee products, beverages, fruits and north medical herbs, to camellia seed and tea leaves. Currently, the certified NTFPs are launched onto both national and international markets with certification label.

3.2.4.1 Certified area

By the end of 2020, the certified NTFP area was approximately 5.72 million ha in China, of which 5.65 million ha were state-owned, and located in key forest areas and natural reserves in the 3 northeastern provinces, Hebei and Hunan, taking up 98.78%; 70,000 ha were collective-owned, and located mainly in North China and South China, taking up 1.22%.

3.2.4.2 Geographical location of certified FMUs

By the end of 2020, there were total 58 certified NTFP FMUs from over 40 prefecture–level cities, autonomous prefectures and regions in 15 provinces. Of which, 18, 13 and 8 were from Heilongjiang, Guizhou and Jiangsu respectively.

3.2.4.3 Certified company

By the end of 2020, there were 2 main types of NTFP certifications, NTFP management (FM) certification and NTFP chain of custody (CoC) certification. In total, 55 companies were FM+CoC certified, taking up 94.83% of the total; and 3 were CoC certified, taking up 5.17%. Among the companies, 23 were group certification, mainly collective–owned forest management units producing fruits, tea and medical materials, accounting for 39.65%; and 35 were single certification, mainly state–owned forest management units, food and beverage processors, etc., accounting for 60.34%.

3.2.4.4 Types of certified products

By the end of 2020, there were over 50 types of certified NTFPs in China, including wild vegetables, edible fungi, nuts, tea, bee products, fruits, dry fruits, forest medical herbs, drinks and various by–products, oils/spices, bamboo shoot, etc. (See Table 3). Different types of NTFPs are originated from different geographical locations, reflecting the local characteristics, e.g. Heilongjiang produces edible fungi, wild vegetables, nuts, bee products and forest medical herbs; Shaanxi and Henan produce dried dates and walnuts; Jiangsu, Zhejiang, Guizhou, Yunnan and Fujian in South China produce tea, *Camellia oleifera* and *Gastrodia elata*.

3.2.4.5 Labeling of certified products

By the end of 2020, total 27 FMUs' certified NTFPs have applied for certification label, including oils, bee products, forest fruits, fungi, bamboo shoots, tea, dried fruits, fruit cakes, fruit jams, drinks, fresh fruits, nuts and others. The applicant

Table 3 Certified NTFP Types

Types	Certified products
Fruits	Winter jujube, pear, apple, plum, blueberry, roxburgh rose, soft-seed pomegranate, huckelberry, peach, etc.
Dried fruits	Dried blueberry, Jun jujube, dried cranberry, etc.
Nuts	Walnut, hazel, Korean pine nut, etc.
Edible fungi	Black fungus, *Pholiota nameko*, *Armillaria mellea*, *Ganoderma tsugae*, *Lactarius deliciosus*, *Hericium erinaceus*,etc.
Wild vegetables	Dandelion, day-lily bud, *Sanicula chinensis*, *Aralia elata*, etc.
Tea	Black tea, green tea, Puer tea, etc.
Bee products	Bee product, bee honey, etc.
Forest medical materials	*Gastrodia olata*, Nothern groundcone, *Pinus pumila* nut, *Dondrobium nobile* series, desertliving cistanche, etc.
Drinks and by-products	Blueberry drink and wine, fruit wine, brandy, fruit jam, etc.
Oils/spices	*Camellia oleifera* fruit/camellia oil, rosemary, etc.
Bamboo shoot	Bamboo shoots

companies accounted for 45% of the total certified companies. Since labels were put into use in 2015, the number of issued labels has accumulated to 1.6 million, mainly for tea, fungi and bee products which reached over 300,000.

3.2.5 Certification benefits

Since the NTFP certification pilot was initiated in China, it has made remarkable achievements in promoting sustainable NTFP resources management, ecological conservation, community harmony, economic development and forest area governance. It meets not only the Chinese long-term development strategy, but also the concept of eco-civilization.

3.2.5.1 Economic benefits

Firstly, the NTFP certification promoted the development of local non-timber product economy and generated diversified incomes in a macro sense. It provided more routes for local farmers to make fortune, and injected new life to local economy. Secondly, NTFP certification increased the products' added value, and realized 10%–15% premium. Thirdly, NTFP certification increased the market access and market share for products. Fourthly, the NTFP certification improved the products' competitiveness, facilitated the certified products' access to international market, and boosted international trade. Fifthly, the NTFP certification enables the thorough tracing of certified products. With the unique QR code on each label of the smallest package unit, the certified product can be traced down to its origin. In this way, the traders' and consumers' concerns in purchase can be eliminated, and thus the product sales can be promoted. Meanwhile, the certification label can be recognized as the "ID" of certified products, which will help to crack down on counterfeit products, and maintain the proper order in the certified products market.

3.2.5.2 Social benefits

Firstly, the NTFP certification increased the forest product producers' and managers' awareness of sustainable forest management and green development, and promoted the harmony and development of local community. Secondly, the NTFP certification provided the residents in forest areas with diversified job opportunities, and improved the corporate image and the sense of social responsibilities. In the context that the state-owned forest farms are being reformed, and the commercial logging is banned in the state-owned forest areas in Northeastern China, the NTFP certification played a significant role in securing the employees' rights and interests, providing job opportunities, and safeguarding the forest farm staffs' life. Thirdly, the NTFP certification introduced the concept of sustainable forest management to the whole society, and improved the community's awareness of forest certification and green consumption. As the certified NTFP processors increased in number, more and more label issued products can be seen in market, E–commerce platforms and ordinary families. Consequently, the concept of sustainable forest management will be communicated to the great public, and their awareness of green consumption will be increased.

3.2.5.3 Ecological benefits

Firstly, the NTFP certification increased the environmental protection awareness of staffs in forestry authorities and members of group certification, turned the forest environment protection into the staffs' or forest farmers' conscious acts, and thus improved the forest resources conservation. Secondly, NTFP certification requires the certified FMUs to consider environmental protection and forest eco–function conservation as the criteria in making production and operation decisions, and enables them to play the demonstrative role in promoting local eco–environmental protection. Thirdly, the NTFP certification provides an effective regulatory mechanism for sustainable cultivation and utilization of non–timber product resources, and thus facilitates the conservation and sustainable utilization of resources under forest.

Case 3

NTFP Certification Promoting the Transformation and Development of Forestry Services

Founded in March 1963, Yingchun Forestry Bureau is located at the southern foot of Wandashan Mountain in eastern Heilongjiang, and the core of the National Natural Reserve for Northeastern black bee (*Apis mellifera* ssp.). It manages a 217,000 ha of forest area, where the forest cover is 65.22%. Since the logging ban in natural forest was issued, it started the economic transformation, and turned to cultivation of Northeastern black bee and development of bee products.

Yingchun Forestry Bureau obtained the CFCC FM Certificate and NTFP Certificate on December 8, 2013. Total 5 types of NTFPs were certified, including bee products (bee honey, royal jelly, bee pollen, propolis, etc.), edible fungi (agaric, mushroom, etc.), northern medical herbs (*Schisandra chinensis, Acanthopanax* spp., *Cortex phellodendri,*etc.), wild vegetables (fiddlehead (*Pteridium aquilinum*), *Athyrium multidentatum, Rabdosia sculponeata, Osmunda japonica, Aralia elata, Spuriopimpinnella branchycarpa,* etc.), and nuts (hazel, pine nut, walnut, etc.).

The NTFP certification brought new opportunities for Yingchun in the transformation period, and significant social, economic and ecological benefits:

The NTFP certification promoted the development of non-timber product economy, diversified the income sources, and provided the residents in forest area with more opportunities to make money. The bee-keeping households increased from less than 100 to 355, swarms increased from less than 10,000 to 35,000, and the bee-keeping households' income increased from RMB 9.5 million yuan to RMB 43 million yuan. The honey price also tripled in contrast with the previous rate. By protecting and cultivating the forest resources, and increasing the employees' and local people's income, the NTFP certification helped to create the multiple-win situation.

The NTFP certification improved the competitiveness of Yingchun's certified products, promoted their going global, and increased the international trade volume. Currently, Yingchun's certified honey products became very popular among foreign importers as an ecology-friendly brand. Having been exported to Malaysia, South Africa and Korea, the products realized the growths of 22% in price, and 50% in annual sales profit.

The NTFP certification promoted the conservation of forest resources and ecological environment. Keeping the concept of sustainable forest management and green development in mind, Yingchun strictly regulated the development activities of NTFPs, maintained the balance between people's livelihood and ecological conservation, and enhanced the continuous improvement of eco-environment in the forest area.

Case 4

NTFP Certification for Tea Enhancing the Win-win Collaboration Among Stakeholders

Dongshan Town of Suzhou City, Jiangsu Province is the producing area of the well-known Dongting (Hill) Bi Luo Chun Green Tea. To protect the Dongting (Hill) Bi Luo Chun brand, and to promote its sustainable management, Dongshan Town government, in collaboration with 2 tea cooperatives and 2 factories, took the lead in organizing the CFCC NTFP group certification in 2016, and successfully obtained the NTFP certification in September, 2016. Currently, the certification has gradually demonstrated the benefits, e.g. the main stakeholders along the tea supply chain, including tea farmers, workers, cooperatives, processors and dealers closely worked together in the process, shared the benefits of certification, and realized the mutually beneficial win-win situation.

1. Tea farmers

● Expanded the channels to acquire technical and market information, and established more favorable connection with relevant government departments;

● Acquired more opportunities to join tea cultivation trainings, and improved their capacity;

● Paid more attention to ecological and environmental protection in tea farm management, e.g. avoided burning for soil preparation, and prevented water and soil losses; Gradually reduced the use of chemical fertilizers and prioritized organic fertilizers, avoided the banned pesticides for pest control and adopted a comprehensive prevention and control system which focuses on physical control, and protected the biodiversity in tea farms.

2. Workers

● Workers were provided with accident insurance and additional welfares, including free physical examination, labor protection appliance, improved accommodation and upgraded fire protection equipment, to ensure safe operation;

● Got more opportunities to take part in decision making, which greatly improved their sense of responsibility as masters of farm.

3. Tea cooperatives

● Intensified their impacts, and gained high prestige among tea farmers;

● Enhanced the cohesion inside the organization.

4. Tea processors and dealers

● Developed the high-end customer groups, and increased the tea products' prices by 15% in average;

● Improved the enterprises' brand benefits, corporate image and sense of social responsibility;

● Improved the connection with workers, and enhanced the cohesion inside the enterprise;

● Actively participated in the local community's infrastructure construction, provided local residents with jobs and other services, and thus promoted the community harmony.

3.3 Chain of Custody (CoC) Certification

3.3.1 Overview of CoC certification

Wood industry is a sector of processing wood products with wood as the raw material. Its main products include log, sawn timber, chip, preservative–treated wood, fire–retardant wood, wood–based panel, wood flooring, wooden decorative material, wooden furniture, wooden door and window, wooden stairs, wood–structure building, wood craft (wood carving, etc.), wood–plastic composite, stone–wood–plastic material and products, other wood products, etc. As a leader in wood industry in the world, China produced the most wood–based panels, wood flooring and wood furniture. In 2019, China produced 300 million m^3 wood–based panel, ranking first in terms of total production; the large–scale enterprises produced total 891 million m^2 wood flooring, including 425 million m^2 wood–bamboo flooring, 396 million m^2 stone–wood–plastic flooring, and 70 million m^2 wood–plastic flooring; and the wood furniture industry realized a total production of 316 million pieces of furniture.

The CoC certification aims to track the certified products down to forest along the supply chain, in order to ensure the products purchased by consumers are traceable and from sustainably–managed forest, minimize the risk of using raw materials from disputable origins, and provide consumers with clear proofs of raw material origin. Products carrying CFCC label indicate "the raw materials in the products are from CFCC certified sustainably–managed forest or controlled resources."

CFCC CoC certification is applicable to enterprises producing, processing and selling wood and paper products in the wood and paper–making industries, including producer, processor or trader. Certification helps enterprises' production and trade to meet corresponding laws and regulations, in order to get access to the global market with increasing demands, and provide timber forest products and paper products from responsible sources.

3.3.2 CoC certification standard

The CoC certification standard serves as the foundation for China's CoC certification. CFCC CoC certification adopts the *Chain of Custody*(GB/T 28952–2018) as its standard which was issued in 2018. This *Standard* provides the requirements for forest products' chain of custody, and is applicable to the conformity assessment for a third–party sustainable product certification which implemented the CFCC requirements. Considering the Chinese national conditions and practices, some provisions of the *Standard* were revised in compliance with the PEFC *Chain of Custody of Forest and Tree Based Products - Requirements* , to ensure its applicability and operability.

3.3.2.1 Structure

The *Standard* consists of 9 chapters, scope, normative reference, terminology and definition, determination of material types, minimum due diligence requirements, chain of custody method, product sales and information transfer, minimum requirements for management system, laborer rights and interests & social and healthy and safety requirements along the chain of custody, and 3 normative appendixes, CFCC specification for declaration, CFCC logo and multi–site enterprises' implementation of chain of custody.

3.3.2.2 Determination of material type

The *Standard* provides the requirements for determination of material type, mainly in the purchase and supplier selection stages. It requires that for each purchase, the buyer shall acquire necessary information from the supplier, and categorize the purchased materials into certified, neutral and other material types with the *CoC Declaration.*

3.3.2.3 Minimum due diligence (DDS) requirements

The minimum DDS requirements is the most important chapter in the *Standard*, and consists of 6 parts, namely basic requirements, information collection, risk assessment, suggestions or complaints, major risky supply management and market ban. Enterprises shall use the due diligence system for risk management, to eliminate the risk of purchased materials from controversial sources, and collect identification information as the species and harvest place of materials or products from suppliers.

Enterprises shall conduct risk assessment for all forest-based materials covered by DDS, to determine whether they are from controversial sources, and then categorize the raw materials into "very low risk" and "major risk" on the basis of the results of risk assessment. Enterprises shall assess the risks of each supplier in the first purchase, and shall reassess the risks at least once a year. In case of any changes in information collected, enterprises shall conduct risk assessment for each supply, and reassess the risks of supplies concerned when necessary. Enterprises shall ask the suppliers with "major risks" to provide additional information and evidences, to make sure their supplies are of "very low risk." Enterprises cannot include timbers or timber products from unknown or controversial or even illegal sources into their product portfolios covered by CFCC chain of custody, and such products shall be banned from market.

3.3.2.4 Chain of custody method

The *Standard* specifies 2 methods, physical separation method and percentage method for the implementation of chain of custody certification. Enterprises can select the most appropriate method according to raw material and processing method. For enterprises which can ensure certified materials or products never being mixed with those uncertified ones, or certified materials or products always being separated from those uncertified in the whole process, the physical separation method is prioritized. Enterprises shall ensure that certified materials are separated from the uncertified or can be clearly distinguished in the whole process of production or trade. While the percentage method is applicable to the enterprises in which certified materials or products are always mixed with the uncertified, and they should use simple percentage or rolling percentage to calculate the certified percentage.

3.3.2.5 Product sales and information transfer

Product sales and information transfer includes relevant contents concerning product sales, transportation documentation and label use.

3.3.2.6 Minimum requirements for management system

The minimum requirements for management system includes the basic requirements, responsibilities and right, CoC procedural documents, record storage, resource management, inspection and control, complaints and subcontracting.

3.3.2.7 Laborers' rights and interests, social, health and safety requirements along the chain of custody

Laborers' rights and interests, social, health and safety requirements along the chain of custody provides the applicable users and requirements to them, and demonstrates the concern over the laborers' rights and interests. It not only secures the workers' rights and interests, but also reflects the *Standard's* social benefits.

3.3.3 Status of CoC certification

3.3.3.1 Product types of the certified companies

According to the statistics of www.cfcc.org.cn, total 257 China's enterprises and organizations obtained CoC certificate by the end of June 2021, including 85 pulp and paper companies, 22 log and sawn timber companies, 28 wood–based panel companies, 51 wooden flooring and furniture companies, 54 non–timber forest product processing companies, and 17 other companies.

3.3.3.2 Geographical distribution of certified companies

The organizations obtained CoC certificate located in 18 provinces, 3 municipalities and Hongkong, of which Jiangsu Province accommodates the most organizations, followed by Zhejiang Province and Heilongjiang Province. By administrative zoning, 7 were from North China, 40 were from Northeast China, 155 were from East China, 34 were from South China, 8 were from Central China, 7 were from Southwest China, and 6 were from Hongkong.

3.3.4 CoC certification benefits

3.3.4.1 In promoting the standardized development

CoC certification can help enterprises to further complete their management system, ensure the legality and sustainability of material source, realize the traceability of the whole production process, standardize their operation, and improve their management. CoC certification is significant in improving the enterprises' brand building, corporate image, performance and competitiveness. For example, Guangdong Yaodonghua Decorative Materials Co., Ltd., a famous veneered panel manufacturer in China, through applying for the CoC certificate, strictly controlled its purchase, production and sales, standardized its operation activities. In this way, it significantly improved its management, effectively ensured the quality of its core products, improved its "KAPOK" brand image, and boosted the green development.

3.3.4.2 Promoting the development of market

The procurement documents of the Organizing Committee for the 2022 Olympic and Paralympic Winter Games, *Technical Criteria for Sustainable Procurement－Paper Products* and *Technical Criteria for Sustainable Procurement－Wood Products*, clearly provide that only products which satisfied the requirements of *CFCC*

Forest Management and *CFCC Chain of Custody*, and obtained CoC certificate can be included in the purchase list. Another successful case came from "Tiantan" furniture, the downstream product of Tiantan Wood, with which the company had been selected as the "official supplier of living furniture for the 2022 Olympic and Paralympic Winter Games." CFCC created a favorable market environment through the government procurement policy, promoted its CoC certification through the market, and helped the Organizing Committee of 2022 Olympic and Paralympic Winter Games to shape the event's brand culture. Having obtained the CoC certification, the Nature Home can use both the CFCC and PEFC logo. The wood products with the certification logo are well accepted by market, and the EU market. In this way, it helps enterprises to get access to the EU market, and enhance their competitiveness on the international market.

3.3.4.3 CoC certification as the foundation for the national green product assessment, and as a guidance towards green development

On July 1, 2018, the first series of certification standards were implemented for the national green product assessment. The GB/T 35601 – 2017 *Green Product Assessment Wood-based Panel and Wooden Flooring* and the GB/T 35607 – 2017 *Green Product Assessment Furniture* clearly requires that timber raw materials must meet the requirements of GB/T 28951 *CFCC Forest Management or* GB/T 28952 *CFCC Chain of Custody*. The national green product certification can only be applied after the forest certification was obtained. At present, over 20 enterprises, including Dare Global, Tiantan Wood, Power Dekor, Furen Panel, Jiusheng Wood, SunYard Wood, etc., have obtained CoC certificate and the national green product certification. As the national green product certification advances, the enterprises in the wood industry built the awareness of green development. They actively applied for CoC certificate, and effectively promoted the green and sustainable development of China forest industry.

Case 5

Forest Certification Promoting Senmao's Green and Sustainable Development

Jiangsu Senmao Bamboo and Wood Industry Co.,Ltd. (Hereinafter referred to as "Senmao") is an exporter of engineered wood flooring. Located in Yixing, Jiangsu Province, it covers an area of 66,000 m^2, including the 48,000 m^2 workshops. It has total 220 workers and managerial staffs, of which 40 are technical staffs.

In 2017, it started the CoC certification works. In August 2018, it obtained the PEFC/CFCC–CoC certificate for its engineered wood flooring, archaized flooring and veneer laminated wood flooring. Since it applied for the CoC certificate, Senmao continuously improved its management system as required, and realized the precise traceability of products. It actively promoted the certified materials. With the CFCC/PEFC logo on its wood flooring products, Senmao image was significantly elevated, and highly praised by consumers both at home and abroad, which generated considerable sales volume for the company. The CoC certification has brought the following 3 benefits for Senmao:

● Optimized internal management

Forest certification helped Senmao to build a complete forest certification management system, a whole–process supervision mechanism and a standardized tracing system, to ensure the legal and sustainable material sources, realize the whole–process traceability, standardize the operation activities, and improve the management.

● Improved corporate and brand images

Forest certification can be used to prove the responsible purchase, get access to the green sales channel for forest products, increase the customers' and consumers' confidence and trust, and improve the company's credibility and brand awareness. The CoC certification is significant for an enterprise's brand building, corporate image, performance and competitiveness.

● Increased domestic and international market shares

Forest certification helps to increase the domestic and international market share, attract quality buyers both at home and abroad, and provides the opportunities to get access to the international market. Having obtained the CoC certificate, Senmao could get into the EU market more easily with the CFCC and PEFC logo. Since it got the forest certification in 2018, Senmao expanded its export to Canada, Australia and Europe from the previous single overseas market—the U.S., and thus increased its production and sales by 30%, and market shares by 15%. When the China-America trade faces extreme difficulties, it quickly opened the European and South American markets with the certified wood products, and maintained the stable and sustained development with the total 12 million USD orders.

Senmao attaches great importance to eco-civilization. Aiming at accelerating the industry's green development, it actively promotes the sustainable flooring industry, and provides the consumers with quality and reliable products and services. It will continue to strictly practice the concept and purpose of forest certification, implement the international strategy, further expand the domestic and overseas market share, accelerate the brand building, promote the green wood and bamboo products, and facilitate the green development, transformation and upgrade of wood and bamboo industries. Senmao will take concrete actions to implement its commitment—Protect the green mountains and lucid water with forest certification!

3.3.5 Problems and challenges in the CoC certification

The CoC certification faces the following 4 problems and challenges in China:

3.3.5.1 Enterprises' insufficient capacity in CoC management

Enterprises can hardly find the appropriate suppliers when factors as tree species, price and risks are considered in purchase. The risk assessment for raw materials in products must be conducted to determine whether the sources are controversial. But most of enterprises have little knowledge of how to conduct risk assessment for materials. Usually in production, they tended to fail in physically separating the certified materials from the uncertified in warehouses, transition areas and on product lines.

3.3.5.2 Insufficient supply of certified materials

China lacks of certified wood materials, especially the commonly used species like poplar, eucalyptus from plantation and bamboo materials. Also it lacks the certified material manufacturers, especially the plywood and decorative based paper manufacturers. The insufficient supply of semi-finished products and broken supply chain jointly led to the difficulty in promoting the CoC certification extensively among timber processors and manufacturers.

3.3.5.3 Insufficient capacity building

Forestry authorities, FMUs, processing and manufacturing enterprises, and the public have little awareness to forest certification, together with enterprises' managements and internal staffs responsible for certification works lack of systematic knowledge of CoC certification standard and operation, jointly led to the difficulty in the enterprises' building of management system.

3.4 Bamboo forest certification

3.4.1 Overview

Bamboo is one of the important forest resources in China. Known as the "green gold mineral," bamboo resource is characterized by wide coverage, fast growth, extensive applicability, high adaptability, and high ecological and economic value. China is rich in bamboo resources, including the sympodial bamboo in tropics, monopodial bamboo in subtropics, and mixed bamboo in high-altitude and high-latitude areas. As one of the main producing areas of bamboo, China has 1/5 of the bamboo forests in the world. It has the richest bamboo resources, the largest bamboo forest area and production, the longest cultivation history, and the highest management level in the world. As the largest producer country of bamboo products, China ensures an extremely promising market prospect for enterprises.

3.4.1.1 Significance of bamboo forest certification in China

Bamboo forest certification provides the bamboo product producers, traders and consumers an environmentally and socially responsible solution, and an effective means for long-term bamboo forest resource management. It can ensure the bamboo forests' long-term health and high productivity in the long run, and thus enable the bamboo forests' functions in bamboo timber production, wildlife habitat conservation and water source cleansing, while securing its social benefits.

3.4.1.2 Targets of China's bamboo forest certification

According to the 9th national forest resources inventory of China, there were total 6.41 million ha bamboo forests in China, including 4.68 million ha of *Phyllostachys pubescens* bamboo forests, and 1.73 million ha other bamboo forests. In 2018, the bamboo industry realized a production value of RMB 245.6 billion yuan. Considering its unique characteristics, bamboo forests can be categorized into 3 types by purpose, i.e. bamboo forest for timber use, bamboo shoot use, and both timber and bamboo shoot use.

3.4.1.3 Relationship between bamboo forest certification and forest certification

Bamboo forest has the unique characteristics of fast regeneration and high yield. Bamboo can grow dozens of meters within a year, and produce timbers in 3–5 years, while trees usually have a long growth cycle, and require dozens of or even hundreds of years to produce timbers. The unique characteristic of bamboo forest provides the advantage of short recovering period of certification, which may increase the certification holders' economic benefits. Additionally, the bamboo forest certification standard is much simpler than the forest certification standard, and thus many assessment procedures can be simplified.

3.4.2 Certification standards

Bamboo forest management certification standard is based on forest management certification standard. It is aligned with the international forest certification standard, and takes care of the unique characteristics of bamboo forest management. The certification indicators for bamboo forest considered and combined the developments of economy, society and environment.

China's bamboo forest management certification adopts the standard of *CFCC Bamboo Forest Management* (LY/T 2275–2014), which was officially published on August 21, 2014 and implemented on December 1, 2014. The *Standard* specifies the indicator system which shall be adopted by the bamboo forest management certification, and the performance requirements for FMUs. The bamboo forest management certification standard provides a basis for reviewing and assessing the bamboo forest management in China. It is applicable to the assessments for bamboo forests for timber and fiber production as well as bamboo shoot and timber production. The *Standard* covers 9 aspects, namely legal obligations, bamboo forest ownership, local community and laborers' rights and interests, bamboo forest management scheme, bamboo forest resource cultivation and utilization, bamboo forest protection, environmental impacts, forest monitoring and documentation management.

3.4.2.1 Laws and policies

This part touches upon 2 aspects, i.e. laws, regulations and international conventions as well as bamboo forest ownership. It requires that bamboo forests must be legally managed in accordance with the existing national laws and regulations as well as international conventions, the ownership must be clearly defined, and due obligations and responsibilities must be fulfilled.

3.4.2.2 Sustainable production

This part touches upon 2 aspects, i.e. forest management systems as well as bamboo forest resources cultivation and utilization. It requires that the input and output must be fully considered in bamboo forest management, the rational management system must be developed, training and capacity building must be provided, and diversified operations, local processing and utilization as well as waste reduction must be encouraged.

3.4.2.3 Eco-environmental protection

This part touches upon 3 aspects, i.e. biodiversity conservation, environmental impacts and bamboo forest protection. It requires strengthening the protection of rare, precious and endangered fauna and flora species, environmental impact analysis before management, reduction of environmental damages by multiple means, and healthy bamboo forest management.

3.4.2.4 Public benefits

This part touches upon local community and laborers' rights and interests. It requires fully considering the impacts on local residents, ensuring not to damage local residents' rights and interests, and building a favorable relationship with local community.

3.4.2.5 Monitoring and archive management

This part touches upon the bamboo forest monitoring system, documentation management and bamboo timber tracking & management system. It requires bamboo forest management units to build a monitoring system to regularly monitor the bamboo forest management, and systematically manage the documentation and records.

3.4.3 Status of certification

Bamboo forest certification developed rapidly in China in the past decade. At present, there are 13 CFCC certified bamboo forest management units, of which 4 obtained the bamboo forest management certificate, i.e. Changning County State-owned Forest Farm in Sichuan Province with 1,988 ha certified bamboo forest, Hejiang County Fubao State-owned Forest Farm with 1,333.34 ha certified bamboo forest, Anji Bamboo Industry Association in Zhejiang Province with 3,802.66 ha certified bamboo forest, and Jiangsu with 333.34 ha certified bamboo forest. Meanwhile, Chibi Two-mountain Farmers' Cooperative in Hubei and Wan'an County Wuyun Forest Farm in Jiangxi are applying for the certification for their respectively 8,000 ha and 333.3 ha bamboo forests. Another 9 bamboo products manufacturing enterprises in Zhejiang, Sichuan, Hubei and Jiangsu have obtained the CoC certificate.

3.4.4 Certification benefits

3.4.4.1 For bamboo forest management

Bamboo forest certification helps to improve operation and management, effectively protect and utilize bamboo forest resources, increase production and quality, lift up bamboo products' premium, and create a stable bamboo forest eco-environment for communities in the long run.

3.4.4.2 For bamboo forest managers

Bamboo forest certification generates better protection of local environment, standardized management, job opportunities, and actions to realize the sustainable management goals.

3.4.4.3 For bamboo product dealers

Bamboo forest certification can improve the corporate image, help enterprises to take the initiative in international trade, and increase their products' competitiveness on both domestic and international market.

3.4.4.4 For bamboo product consumers

Bamboo forest certification can raise the consumers' awareness of green consumption, help them to identify bamboo products from responsible sources, inform them of sustainable development information and policies, and encourage them to take actions towards sustainable development.

Case 6

Bamboo Forest Certification in Anji

Anji County in Zhejiang Province is rich in bamboo forest resources, so its bamboo industry developed rapidly, and became a highly developed demonstrative industry in China. As a saying goes, "Zhejiang is the major pillar of the bamboo industry in China, and Anji is the major contributor for the bamboo industry in Zhejiang." In recent years, Anji, in developing its bamboo industry, faced the challenges of bamboo timber price decrease, production and labor costs increase, and failure in product export. By relying on the Bamboo Industry Association, Anji applied the group certification involving the "Association, Village committee and Farmers" in 2020, and obtained the CFCC bamboo forest certificate for its 3,802.66 ha bamboo forest, in order to promote quality development of bamboo industry, and solve the problems in material price and export.

Bamboo forest certification generates the following benefits:

1. It meets both the domestic and international market demands, and expanded the sales channel of bamboo timber.

2. Certified bamboo timber price is higher than the uncertified, and the added value of bamboo products is improved.

3. It provides access to the supply side of certified materials, and set the direction of future sustainable development for the bamboo industry in Anji.

4. It laid a solid foundation for breaking the international trade barrier and supplying for national government procurement, and played an effective role as a radiator and driver in developing the neighboring bamboo producing areas' bamboo industry.

3.5 Eco-system service certification

Different natural protected areas are important in providing ecosystem services and quality ecological products. The *Guiding Opinions on Building the National Park-based Natural Protected Area System* was published in 2019, which provides a policy guarantee for better promoting the implementation of ecosystem service certification.

3.5.1 Overview

Ecosystem service certification is to assess the eco–environment service functions in natural reserve, forest park, etc. It is to inform the public of the potential value of different forest eco–environment services, further advance the sustainable protection and management of forest on which eco–environmental services depend, and balance the development, utilization and environmental protection. In this way, the eco–environment services of forest resources can be fully developed, while the eco–environment is effectively protected, and the eco–products' value can be realized.

Forest eco–environment service certification is a kind of forest management certification. It refers to the third–party certification body's independent audit and assessment for the FMUs offering the forest eco–environment services in accordance with relevant certification standard, to verify whether it meets the sustainable management requirements, i.e. the continuous and harmonious co–existence of eco–environmental protection and development. The certified FMUs should meet the requirements of forest management certification standard when applying for the forest eco–environment service certification. At present, the main targets of eco–environment service certification in China's forest certification scheme are forest park, natural reserve and other natural protected area. As China's forest certification scheme continuously increases its scope, the forest eco–environment service certification will get further development in the future.

3.5.2 Certification standard

By the end of 2020, China has published and implemented 4 ecosystem service certification standards, all of which are forestry standards for natural protected area. The standards cover 4 types : forest park, natural reserve, forest health preservation and eco–tourism in natural protected area. Of which, the natural reserve and forest park eco–environment service certification standards were incorporated into the *Green Bond Endorsed Projects Catelogue* (2021), as the condition for supporting the protection-oriented projects in national parks, world heritage sites, national scenic spots, national forest parks, national geological parks, national wetland parks, etc.

3.5.2.1 Natural reserve eco-environment service certification standard

The *Eco-environment Services in Nature Reserve*(LY/T 2239-2013)includes 7 principles, 35 criteria, and 108 indicators. It covers 4 main contents:

① Basic requirements: This part covers 3 aspects, namely legal obligations, operation and management, and community management. It requires management units to manage natural reserves in accordance with the existing national laws and regulations, international conventions, and bi-/multi-lateral agreements, develop a scientific and feasible natural reserve plan, organize sufficient mechanisms, personnel, funds and facilities, and ensure employees' safety, rights and interests.

② Biodiversity protection: This part covers the zoning plan, control of human activities, field patrol inspection and protection, key wildlife protection, ban of alien species, and protection of genetic diversity in natural reserves.

③ Eco-environment: This part covers the control and management of air, surface water, solid waste, noise and light pollution, and soil erosion in natural reserves.

④ Natural reserve management: This part covers 2 aspects, scientific research, publicity and education as well as eco-tourism. It requires management units to scientifically measure biodiversity, test eco-products in a standardized manner, implement scientific research, publicity and education, develop eco-tourism, and regulate tourism infrastructure and commercial activities, in order to ensure tourists' safety and security.

3.5.2.2 Forest park eco-environment service certification standard

The *Eco-environment Service in Forest Park*(LY/T 2277-2014) covers 6 principles, 32 criteria, and 101 indicators. It consists of 4 parts:

① Forest landscape resources and biodiversity conservation: It covers the lawful and orderly planning and construction, protection of wildlife and landscape resources in forest parks, appropriate vegetation protection measures, effective protection for rare, endangered and national key protected wildlife, control of alien species, resource inventory and monitoring, etc.

② Operation and management of forest park: This part covers 3 aspects, namely eco-culture dissemination, eco-tourism products and services, and eco-environment maintenance. It requires full and complete facilities and systems for scientific education service, diversified channels and rich contents for eco-culture publicity, various eco-tourism products, complete facilities, and effective control over ecological and environmental pollutions.

③ Regional economic development: It requires management units to respect the traditional lifestyle of residents in surrounding communities, establish an effective communication mechanism, and promote the regional economic development.

④ Basic requirements: This part covers legal obligation, management system, safety management, employees' rights and interests, etc. It requires complying with laws, regulations and standards, establishing the operation management mechanism and management system, strengthening safety protection, and securing employees' rights and interests.

3.5.2.3 Natural protected area forest wellness certification standard

The *Forest-based Health Preservation in Natural Protected Area* (LY/T 3245-

2020)includes 8 principles, 19 criteria, and 90 indicators. It mainly covers 4 aspects:

① Basic requirements: This part covers 4 aspects, namely legal obligation, ownership, forest wellness operation and management, and community development. It requires the management units to comply with national laws and regulations, carry out forest wellness operation after the land ownership and operation management authority are clarified, legally develop forest wellness, and enhance the community's economic development.

② Development of forest wellness: This part covers the site selection for forest wellness, planning and product option. It requires a location close to forest, easy access to transport means, high forest quality, outstanding environmental conditions, scientific and reasonable forest wellness plan, appropriate product option as well as complete facilities and equipment.

③ Comprehensive services and safety: This part covers 3 aspects, namely people, facility and safety. It requires professional service staffs, complete consulting services and facilities, and safe environment.

④ Eco-environment protection: It requires management units to control air, surface water and solid waste pollutions in accordance with national standards.

3.5.2.4 Natural protected area eco-tourism certification standard

The *Eco-tourism in Natural Protected Area* (LY/T 3246-2020) includes 9 principles, 30 criteria, and 102 indicators. It consists of 4 parts:

① Basic requirements: This part covers 7 aspects, namely legal obligations, ownership, franchised operation, planning, management, community development and tourist satisfaction. It requires management units to operate the eco-tourism and plan for the eco–tourism in natural protected areas according to national laws and regulations, build a complete and appropriate eco–tourism management system, promote the communities' economic development, and ensure the customers' satisfaction.

② Eco–tourism services and facilities: This part covers rest, ecological experience, nature education and requirements for tourism facilities. It requires eco–tourism services to be provided in accordance with the requirements in the plan, adopt high standards to meet high requirements in building eco–tourism facilities, and the tourism facilities built to be safe and comprehensive.

③ Eco–tourism code of conduct and safety: This part covers 2 aspects, namely code of conduct and safety. It requires management units to clearly define the code of conduct for tourists and employees involving in eco–tourism, and properly manage fire prevention, hazardous location, and personal injury first aid management.

④ Eco–environment protection: This part covers 2 aspects, namely ecological and environmental protection. It requires management units to monitor the environmental impacts in developing eco–tourism, avoid tourism routes' running across wild animal habitats and wild vegetation, measure the environmental impacts in a scientific manner, and prevent various pollutions, including air and surface water pollution.

3.5.3 Status of certification

By the end of 2020, China's forest ecosystem service certification area is 7,000 ha mainly distributed in Heilongjiang, Yunnan, Gansu and Hunan. There are 7 management units applied for the forest eco–environment service certification, and 5 of them are in the period of validity. The certified enterprises are mainly national natural reserves, national forest parks and tourism attractions.

3.5.4 Certification benefits

Forest ecosystem service certification has made great achievements in promoting sustainable management, standardized management and ecological conservation of natural protected areas, i.e. the ecosystem service certification can effectively promote the standardized management in natural protected areas, and the realization of eco–products' value. Specifically, the ecosystem service certification can generate the following benefits:

3.5.4.1 Ecological benefits

The ecosystem service certification helped the management units to further improve the eco–environment protection awareness and activities in natural protected areas' management, adopt different measures in establishing natural protected areas, standardize their operation activities, reduce the natural protected areas' eco–environmental impacts, and conserve the eco–functions of natural reserve, forest park and other natural protected areas. For example, in building the infrastructure for Phoenix Hill National Forest Park in Heilongjiang Province, the priority was given to conservation of wild animal habitat and wild plant concentration area.

3.5.4.2 Social benefits

Forest ecosystem service certification provided the following social benefits: Firstly, it promoted the management of natural protected areas, showcased a favorable image, and fully developed the natural protected areas' potential in promoting eco–civilization and quality forestry development; Secondly, it improved the service functions of various natural protected areas, e.g. the Phoenix Hill National Forest Park in Heilongjiang Province added the scientific education and scientific research, e.g. the education and warnings from endangered wildlife, to improve its social value; And thirdly, it strengthened the communication with the residents surrounding the natural protected areas, and promoted the development of regional economy. With the certification, the residents surrounding the natural protected areas, especially the forest parks, have got more job opportunities and higher income.

3.5.4.3 Economic benefits

The ecosystem service certification standardized the management units' management pattern, and improved their image. As the public awareness of forest certification increases, the economic benefits generated will continuously grow in the future.

Case 7

Forest Ecosystem Service Certification Enhancing the Balance Between Natural Reserve Protection and Development

Locating in Beibei District, Shapingba District and Bishan District of Chongqing Municipality, Jinyun Mountain National Natural Reserve is 35 km away from the downtown, and covering an area of 7,200 ha with the forest coverage of 98.7%. With the abundant rare wildlife resources, it is renowned as the typical subtropical ever-green broad-leaved forest and plant species gene bank in the upper-middle reaches of Yangtze River. With its high conservation value and scientific research value, it is regarded as a base for scientific research, ecological teaching and practice as well as environmental education.

In 2020, it finished the major audit for CFCC forest management certification and forest park eco-environment service certification. With the certified area of about 1,200 ha under the management of 10 departments and 6 protection stations under the management bureau, it became the first reserve with CFCC forest eco-environment service certification for natural reserves in southwestern China.

With the natural reserve eco-environment service certification, the Jinyun Mountain National Natural Reserve benefited in the following ways:

● Standardized the eco-tourism in the natural reserve. After certification, it calculated the environment capacity scientifically to prevent overloading.

● Improved the environment protection measures, and intensified the monitoring over air and surface water pollutions.

● Improved and completed the infrastructure. To obtain the certification, it added the eco-tourism products as natural landscape, and reinforced the safety emergency plan and risk prevention measures.

3.6 Wildlife Husbandry and Management Certification

3.6.1 Overview

Wildlife husbandry and management certification is an extension of China's forest certification scheme, and can be used as a supplementary mean for the current regulatory system on the wildlife. It aims to guide and standardize the husbandry and management in artificial wildlife breeding farms as zoo, wildlife park and wild animal feeding farm; and help them to raise the awareness of sustainable management and establish sustainable husbandry and management system.

3.6.2 Certification standard

The State Forestry Administration published the *CFCC Precious and Endangered Wildlife for Production and Management—Feeding and Management* (LY/T 2279-2014) in 2014, and revised it to *CFCC Wildlife Husbandry and Management* (LY/T 2279-2019) in 2019. As an integral part of China's forest certification standard system, the *Standard* is applicable to the artificial wildlife breeding farms' building and implementing wildlife feeding management system, maintaining and continuously improving it to be more appropriate, sufficient and effective. Meanwhile, it is also applicable to the conformity assessment for wildlife husbandry and management undertaken by certification body, since it provides the benchmark to assess whether an artificial wildlife breeding farm has met the laws, regulations and sustainable principles. The *Standard* covers 8 aspects, i.e. general requirements, personnel, facilities and equipment, feeding inputs, husbandry and utilization, animal health, management system documents, inspection and improvement.

3.6.2.1 General requirements

This part provides the basics and coverage of wildlife husbandry and management system, and basic requirements on farm (park) venue.

3.6.2.2 Personnel

This part covers the competency requirements, employees' training, welfare, health and safety, and issues concerning external personnel.

3.6.2.3 Facilities and equipment

This part specifies the requirements for fencing facility, cage and pen, feedstuff storage, processing and preparation, security monitoring facility, disease control, diagnosis and treatment facility in hospital, animal capture, transfer, expelling and fixation, environmental sanitary facility and other equipment.

3.6.2.4 Feeding inputs

This part covers the specific requirements for drinking water, feedstuff, additives and veterinary medicine.

3.6.2.5 Husbandry and utilization

This part covers the animal feeding, nutrition, breeding, products, introduction and transfer, cleaning and disinfection, hazardous biological prevention and control, feeding security and response, etc.

3.6.2.6 Animal health

This part specifies the requirements for veterinary inspection, clinical diagnosis and treatment, anesthesia, ptomatopsia, immunization, veterinary record, zoonosis, animal euthanasia, epidemic response, dead animal body or appendant disposal, etc.

3.6.2.7 Management system documents

This part covers 6 aspects, namely technical documents, management documents, work documents, external documents, animal archive, file and record management, etc.

3.6.2.8 Inspection and improvement

This part covers 3 parts, namely internal audit, traceability inspection and management review.

3.6.3 Status of certification

Total 11 China wildlife husbandry institutions in China have obtained the certification by the end of 2020 (See Table 4). With the husbandry and management system compliant with the requirements of wildlife husbandry and management certification standard, the certificate holders have been playing the active demonstration role in the industry.

Table 4 Types of Certified Wildlife Husbandry Institution

No.	Type of certified institution	Quantity	Type of animal
1	Zoo, wildlife park	2	—
2	Wildlife breeding and rescue center	2	Manchurian tiger, etc.
3	Wildlife farm	3	Turtles and snakes
4	Wildlife-centered scenic spot/forest park	2	Asian elephant, Sika deer, peacock, etc.
5	Wildlife feeding and breeding company	1	Lab monkey
6	Wildlife-centered natural reserve	1	Yangtze alligator

3.6.4 Certification benefits

Forest certification significantly improved the operation and management level of wildlife husbandry institution, and helped them to build the sustainable wildlife husbandry management system. It generates benefits as following:

3.6.4.1 Social benefits

The wildlife husbandry and management certification enabled the wildlife husbandry institutions get the certificates demonstrating their wild animals being from the traceable and sustainable husbandry and management system. Firstly, the certification strengthened the wildlife feeding and breeding farm's awareness of modern sustainable management, built its brand image, and improved its social recognition by intensified publicity. Secondly, the certification provided the scientific and professional third-party certification body's audit and assessment, which could clearly define the corporate responsibilities, reduce the relevant authorities' administrative pressure, and improve work efficiency and social satisfaction.

3.6.4.2 Environmental benefits

Firstly, the environment inside and outside could be significantly improved. The certification standard specified the requirements for infrastructure in feeding environment, layout of farm area as well as design of cage and pen and fencing facilities. Thus each applicant had to complete and optimize the infrastructure inside the feeding farm, and improve the environment, hazard treatment facility, health and medical facilities and other facilities (crushing, processing, preparation, etc.) inside the feedstuff storage and processing plant according to the certification standard and guidance. Secondly, the certification promoted the applicants' introduction and emphasis on professionals, and then the ex–situ conservation, animal pedigree perfection and animal welfare, by which it boosted the progress in wildlife resource conservation.

3.6.4.3 Economic benefits

Firstly, it improved the wildlife feeding and breeding environment, animal population quality and production, so as to secure the economic income. Quality management can directly translate advanced technologies into products, by which it boosted the deep processing and development of wildlife products, increased the added value and technological contents of wildlife and its products, improved the management institutions' competitiveness and brand effect, intensified their soft power, and directly or indirectly increase their revenue. Secondly, the certification could demonstrate the animal resource being from a sustainably managed system, and help to improve the social acceptance and enterprises' interest in production and management. In this way, the win–win situation can be established, ensuring the enlargement of animal population and the sustainable development of resources. Thirdly, the certification could improve an enterprise's comprehensive strength as well as the management and environment inside a wildlife farm, attract more visitors, and generate more income.

Case 8

Wildlife Husbandry and Management Certification Significantly Improved the Husbandry and Management Level of Wildlife Husbandry Institution

Located in Xuanwu District, Nanjing Hongshan Forest Zoo covers an area of 74.82 ha, and accommodates over 3,000 rare animals of 216 species. It obtained the wildlife husbandry and management certificate in September 2016.

With the certification, Nanjing Hongshan Forest Zoo significantly improved its feeding management in the following aspects:

1. Completed and improved the Environment Management Procedures and Facility and Equipment Management Procedures, by adding the control measures for major environment-influencing factors, and standardizing the emergency management in order to prevent accidents in advance.

2. Improved and standardized the processing, storage and health quarantine of animal feedstuff and medication, as well as the storage environment and sanitary condition of feedstuff additives.

3. Developed the specific technical specifications for breeding as required, and added the animal welfare contents to the quality management manual.

Case 9

Wildlife Husbandry and Management Certification Significantly Improved Internal and External Environment

The Wild Elephant Valley is located in the north of Mengyang Town, Jinghong City, Xishuangbanna Dai Autonomous Prefecture of Yunnan Province, and the juncture between the east and west areas of Mengyang Natural Reserve inside

Xishuangbanna Natural Reserve. The Wild Elephant Valley Management Co., Ltd. obtained the wildlife husbandry and management certificate in December 2016 for its 34 elephants feeding and breeding management system.

With the certification, the environment inside and outside the Wild Elephant Valley was significantly improved in the following aspects:

1. The feeding environment and enrichment facilities were greatly improved to eliminate all potential safety and security risks.

2. The feeding inputs' quality control as well as disease and epidemic prevention and control measures were strengthened to effectively prevent digestive and gastrointestinal diseases.

3. The agreement for medical waste disposal was signed with legally-registered and managed medical waste treatment companies, and the disinfecting tank was upgraded, so as to effectively prevent virus and bacteria spread as well as potential sanitary and epidemic risks.

CHAPTER IV
Experiences of Forest Certification Development in China

4.1 Laws, regulations and policies providing the fundamental support for forest certification

The Central Government Document requires "To actively promote forest certification and align with the international principles" in 2003, which laid the legal and policy foundation for forest certification development in China. In the past 2 decades, national laws and regulations as well as the authorities' policies as the *Rules for Forest Certification*, the *Forest Law of the P.R.C.* and the *Green Bond Endorsed Projects Catalogue* took the forest certification into account successively, which greatly intensified the forest certification's role and impacts as both a conformity assessment in the forestry sector, and the important philosophy and basis for the whole society in combating climate change, and promoting ecological conservation and green development. With the great supports from the laws, regulations and policies, the forest certification developed rapidly in China.

4.2 Government support is key to rapid development of forest certification in China

The forest certification's rapid development from weak to strong, from pilot to widely covering, is attributable to the joint efforts by all stakeholders, including the government departments, NGOs, scientific institutions and enterprises. The governmental emphasis is key to the success of China's forest certification scheme. The State Forestry Administration started the top-level design for forest certification in 2001. In the past 2 decades, SFA carried out systematic measures and made major progresses in forest certification system, standard, policy and institution development.

The forest certification improved the FMUs' awareness of sustainable forest management. Both the state-owned forest farms and the forest farmers' cooperatives paid more attention to forest management scheme, biodiversity conservation, employees' rights and interests as well as community development than they did before, in order to meet the certification standards. The certification applicants have gradually formed the awareness of sustainable management, which features the coordinated environmental, social and economic development, and practiced the concept in forest management.

4.3 Multi-stakeholder participation promoting the healthy, rapid and orderly development of forest certification

Forest certification emphasizes the stakeholders' participation at the very beginning, since sustainable forest management considers multi-stakeholders' rights and interests and needs. Only by properly handling different relationships, the harmonious development of human, society and nature can be realized. The National Technical Committee for Sustainable Forest Management and Forest Certification Standardization, which is responsible for developing and managing forest certification standards and technical specifications, consists of stakeholders from government departments, research institutions, industrial associations, education institutions and enterprises. It has been contributing significantly to the standards' development and revision as well as standards' quality and forest certification's impacts improvement.

4.4 CFCC innovatively expanded the scope of forest certification

Foreign forest certifications include forest management certification and chain of custody certification only, and certified products are limited to timbers and timber products. However, China forest certification, considering it's national conditions and forest conditions, innovatively expanded the scope to cover NTFPs, bamboo forest and ecosystem services. In the context of building the eco-civilization, China forest certification emphasizes the multiple functions and benefits of forest, by which it adapted to the development trend in the new era, and expanded the meaning and application on the basis of international forest certifications. It disseminated the concepts of "eco-product" and "green product" among the consumers, and increased the public awareness of green consumption.